The M&M's® Brand Counting Book

Barbara Barbieri McGrath

SCHOLASTIC INC.
New York Toronto London Auckland Sydney
Mexico City New Delhi Hong Kong

The author would like to thank the following people for their assistance and patience, Will M., Roger and Dianne G., Albert B. Jr., Joanne B., Jerry P., John B., Mary Ann S., Sue S., Drew Y., M and D, and Karen S.

This book is dedicated with love to Will, Emily, and W. Louis

—Barbara Barbieri McGrath

ISBN 0-439-23208-2

 Published by Scholastic Inc., 555 Broadway, New York, NY 10012, by arrangement with Charlesbridge Publishing.

12 11 10 4 5/0

Printed in the U.S.A. 08

First Scholastic printing, September 2000

Cover design by Roger Glass

Consultant to the Editor
Ann Foley, 2nd Grade Teacher
Eisenhower Elementary School, Wauwatosa, Wisconsin

The exact number of pieces of each color of "M&M's"® Chocolate Candies may vary from package to package. As a result, there may be insufficient quantities of each color of "M&M's"® Chocolate Candies in each package to match the quantity of colors required for play in this book.

Pour out your candies.
Get ready, get set.
This counting book
Is the tastiest yet!

Call out the colors,
You'll find one that's new . . .

Brown!

Green!

Orange!

Red!

Yellow!

and

Blue!

This certainly is a most colorful mix!
Now sort them as I do and count up to six.

One blue

1

One

Two green

Two

Three orange

Let's go.

3

Three

Four yellow

4

Four

Five red

5

Five

Six brown.

Good show!

Six

Let's count to twelve now.
That's a dozen, you know.

Keep the six brown.

Push the others aside.

Please use these pictures
and words as your guide.

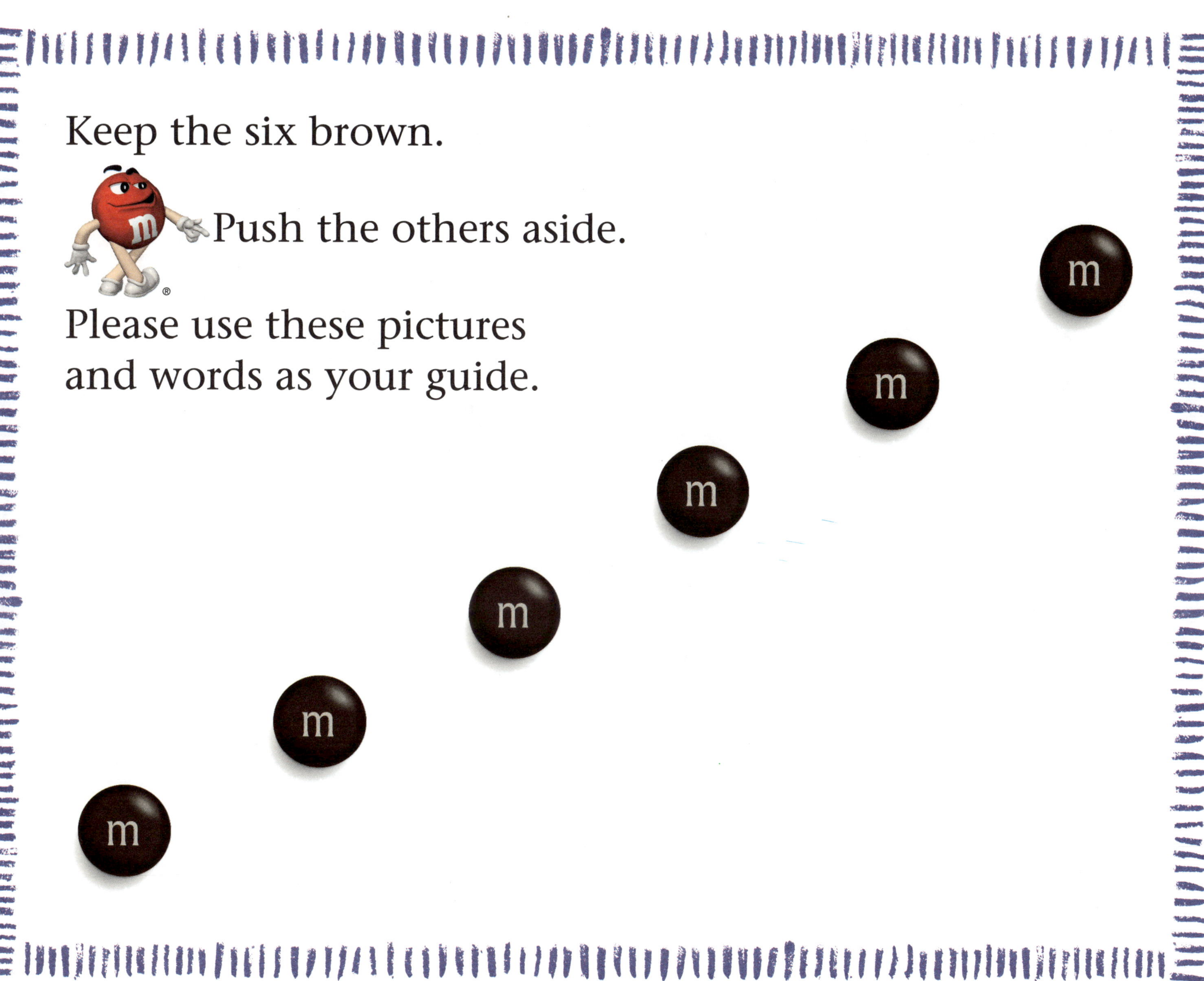

Now add a blue to make seven . . . that's great!

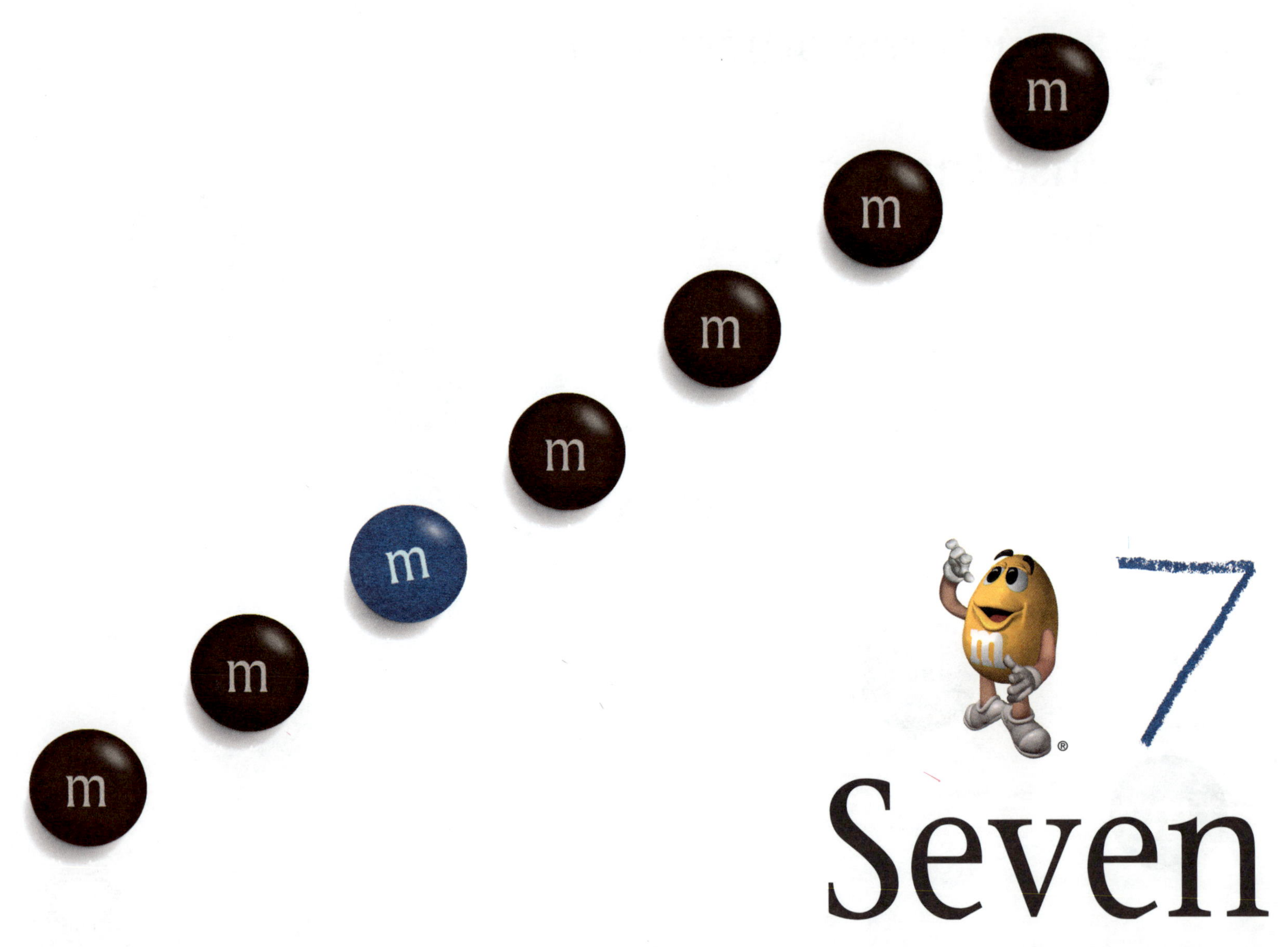

Seven

Put a green near the middle.
Now you have eight!

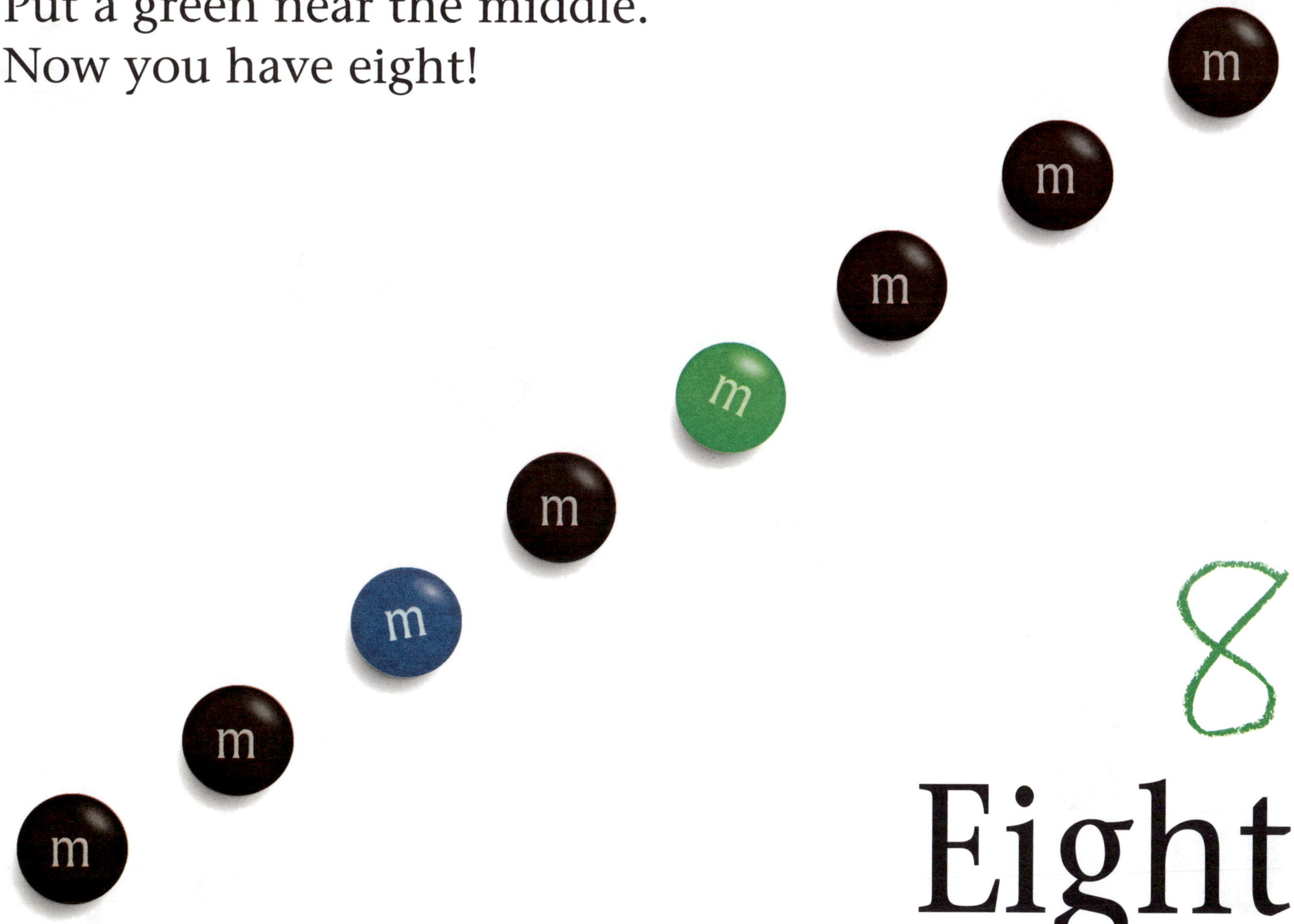

8

Eight

Add a red to your group to make nine, and then . . .

9

Nine

Add the orange and
now we've already reached ten!

10

Ten

To get to eleven, add a yellow one now.

11

Eleven

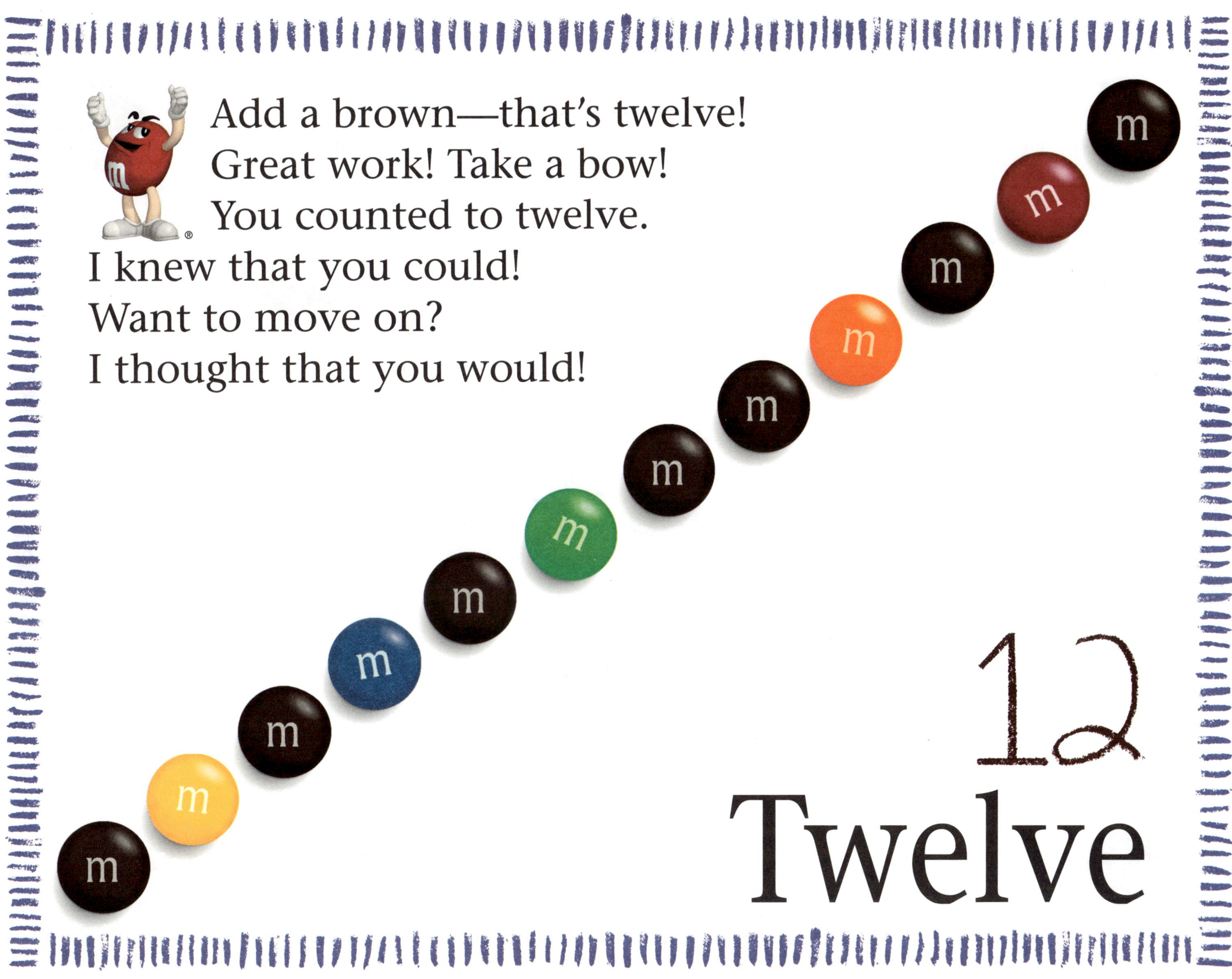

Add a brown—that's twelve!
Great work! Take a bow!
You counted to twelve.
I knew that you could!
Want to move on?
I thought that you would!

12
Twelve

Now put the twelve candies in a long line . . .
We call this a set. You're doing just fine!

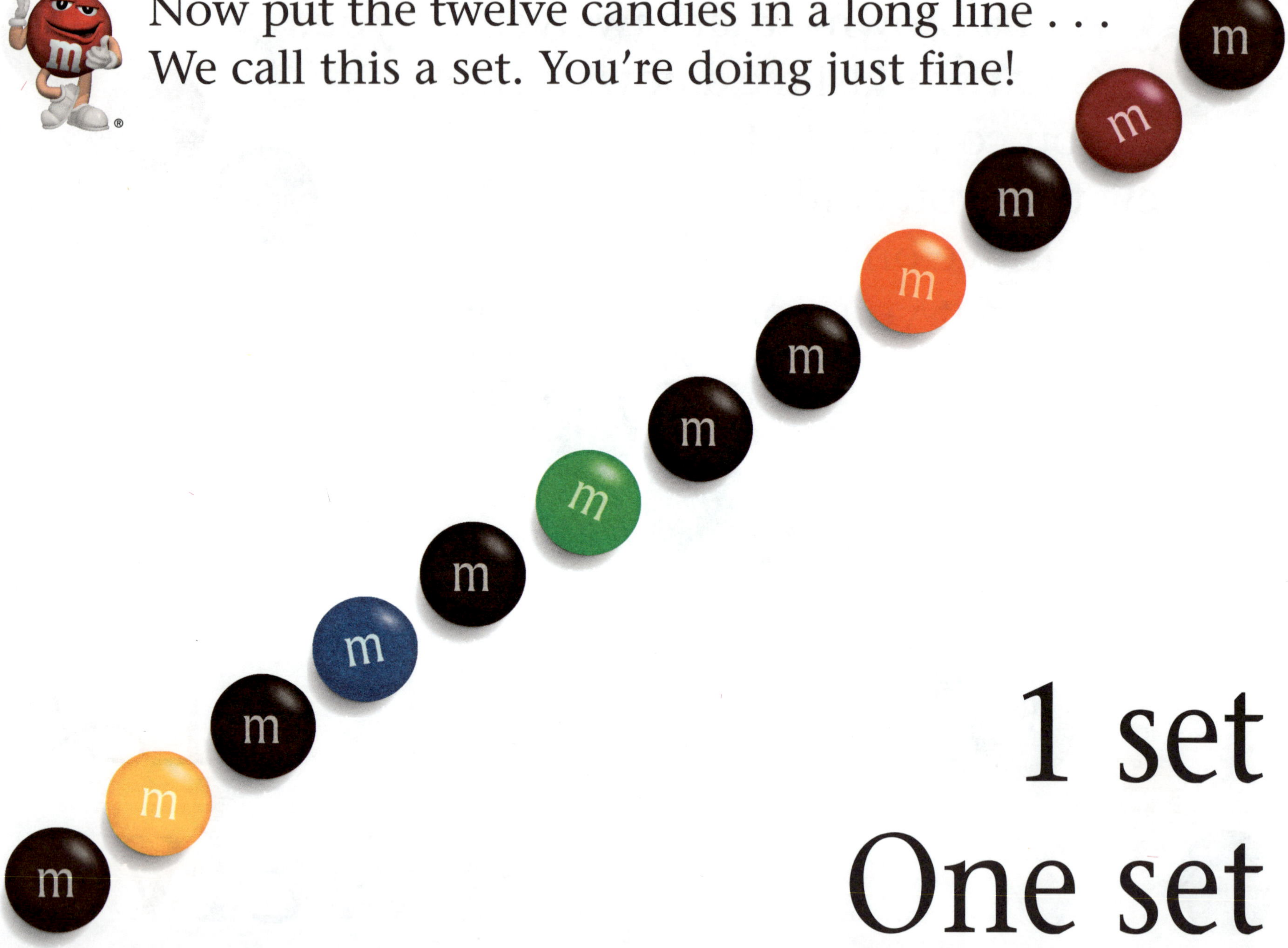

m m m m m m m m m m m m

1 set
One set

Make six groups of two. That's easy for you.
What does this make? Six sets of two!

6 sets

Six sets

Change them around to make three sets of four.
Count them. How many? Still twelve and no more!

4
4
+ 4
12

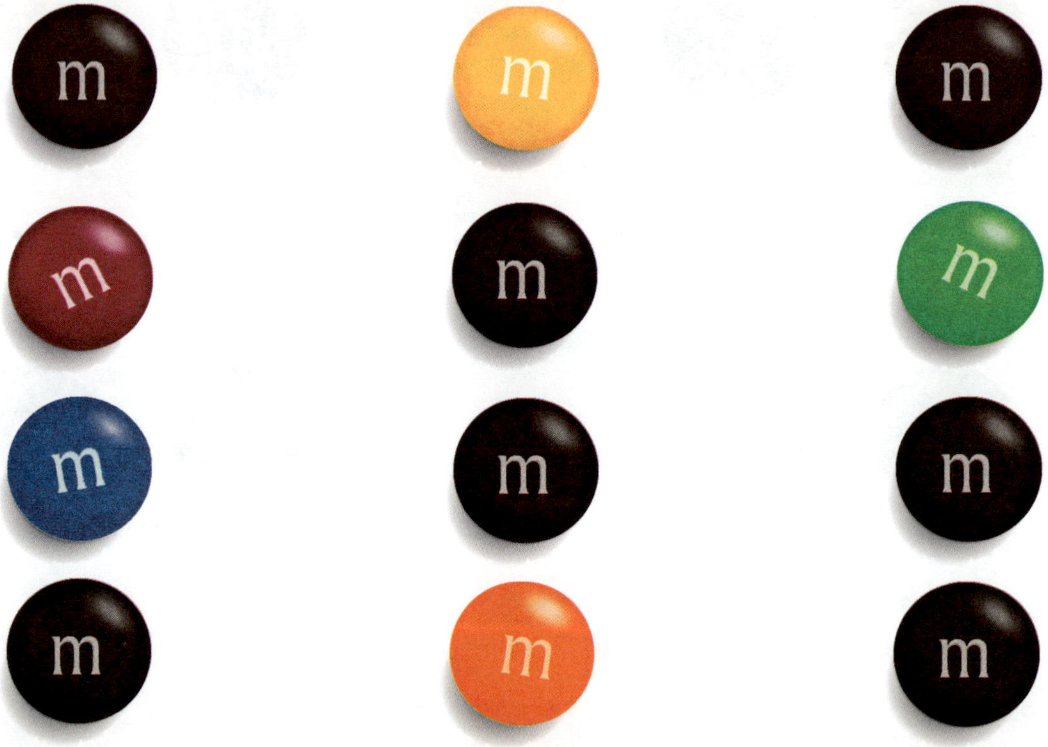

3 sets

Three sets

Now it is time to make four sets of three.
There are still only twelve, as you can see.

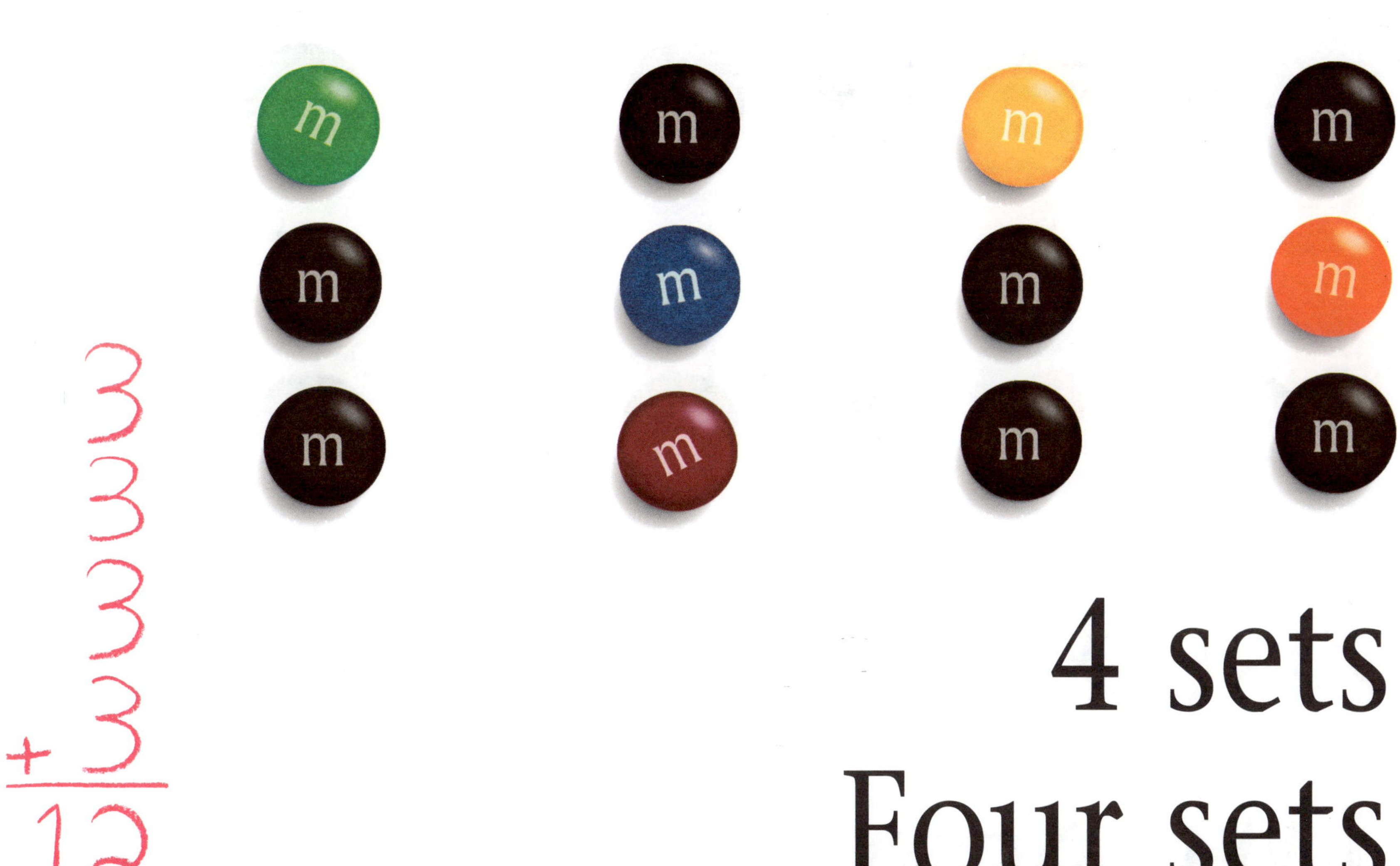

4 sets
Four sets

Make two sets of six now. How did you do?
You did that so well—let's start something new.

2 sets

Two sets

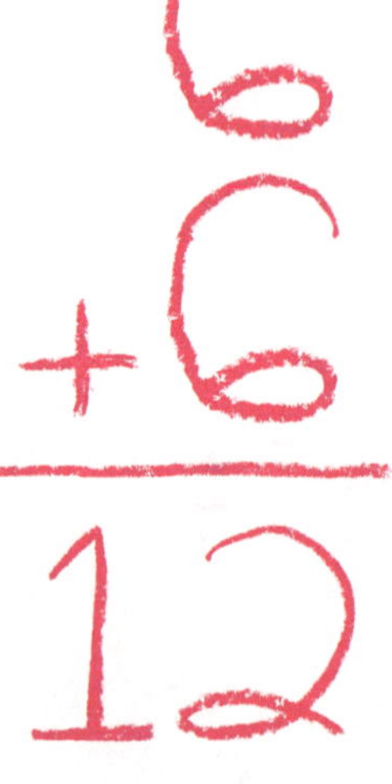

Shape the twelve candies, please, into a square.
A square has four sides. Please count them with care.

Square

Change the square to a circle, the big round kind.
A circle's beginning is so hard to find.

Circle

Let's make a triangle before we stop.
Give it three sides and a point on the top.

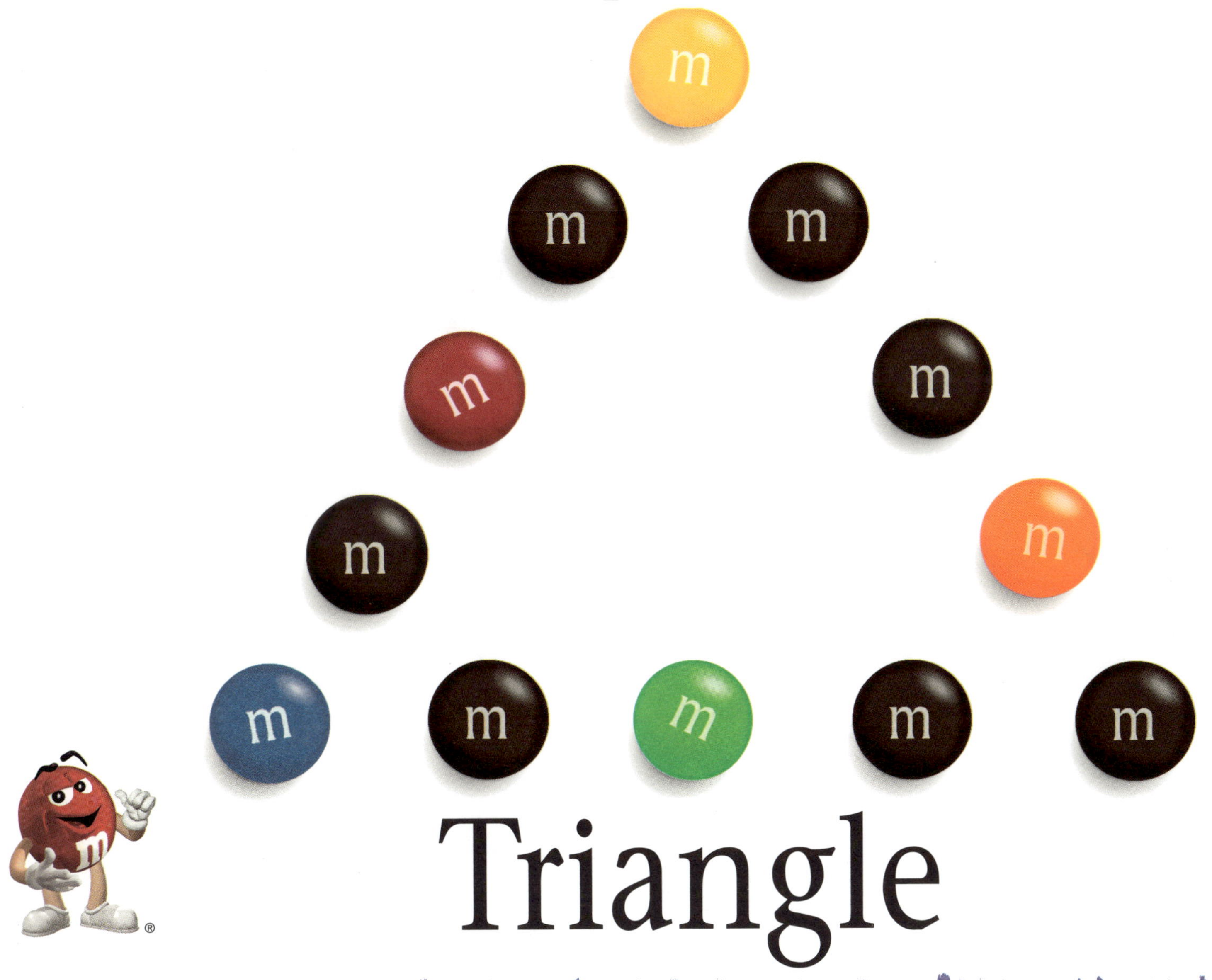

Triangle

Now comes the part that will be the most fun.
We'll start to subtract—
so eat the blue one.

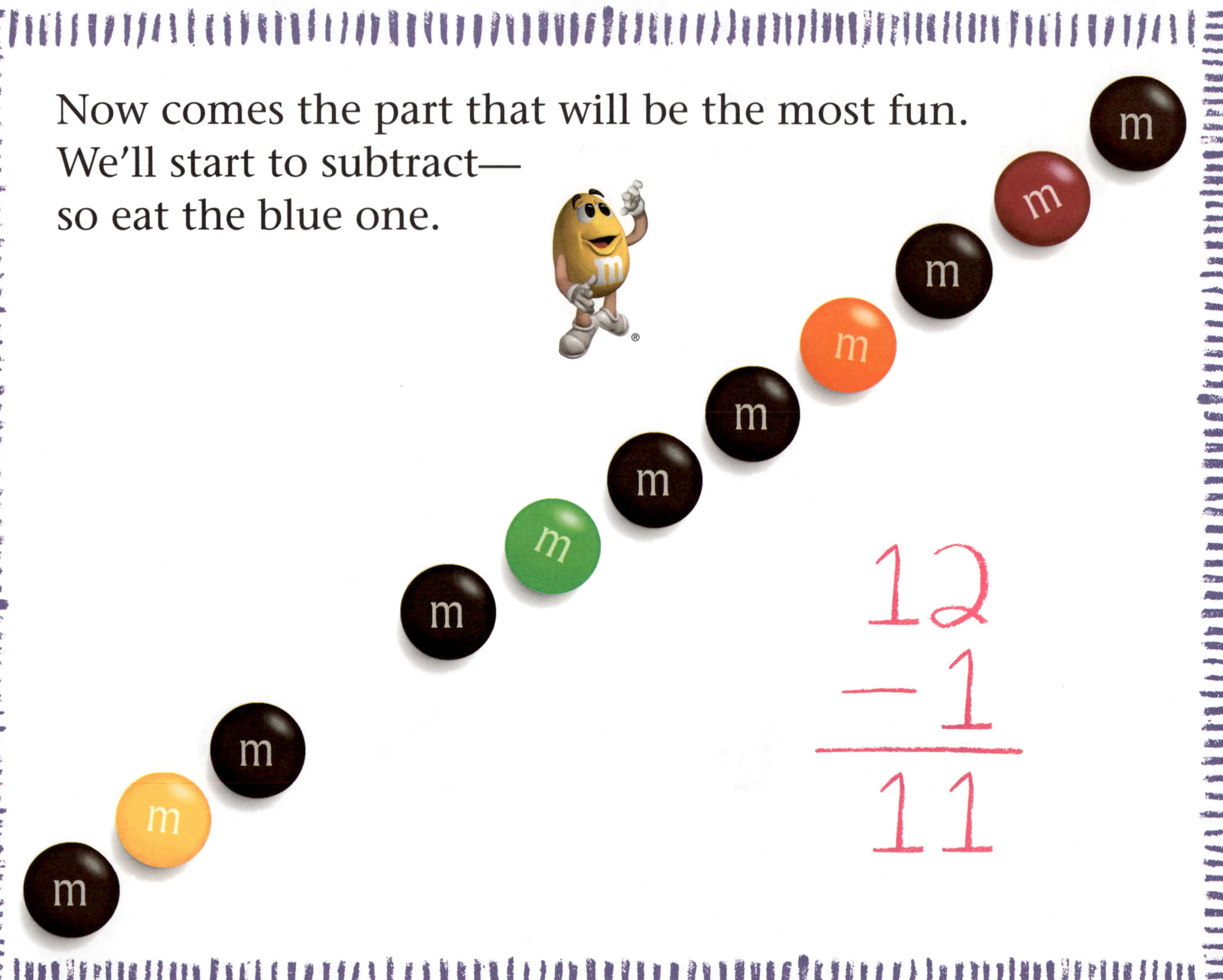

Count and you'll find you now have eleven.
Next eat the brown. Go on, eat all seven!

$$11 - 7 = 4$$

Now you have four left.
So please eat the green.
The orange, red, and yellow
still can be seen.

Next eat the orange one, and after you do . . .
How many are left? Oh dear, only two!

Now eat the yellow—it's second to last.
Counting like this makes time fly by fast.

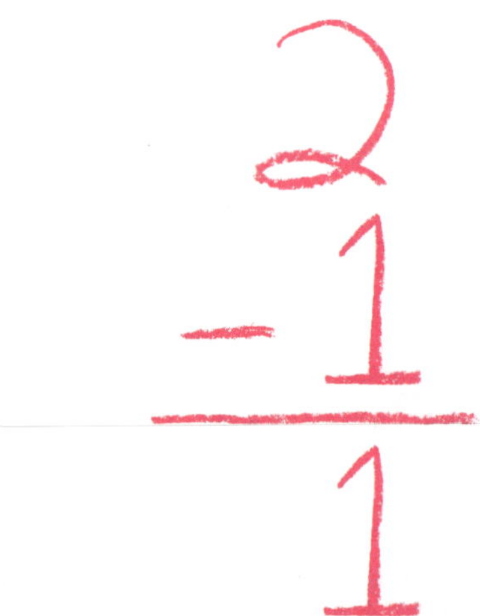

How many are left? You're right. Only one.
Eat the red that is last. Now you are done!

$$\begin{array}{r} 1 \\ -\,1 \\ \hline 0 \end{array}$$

When none are left—it means there are zero.
You've reached the end. You're the grand champion hero!

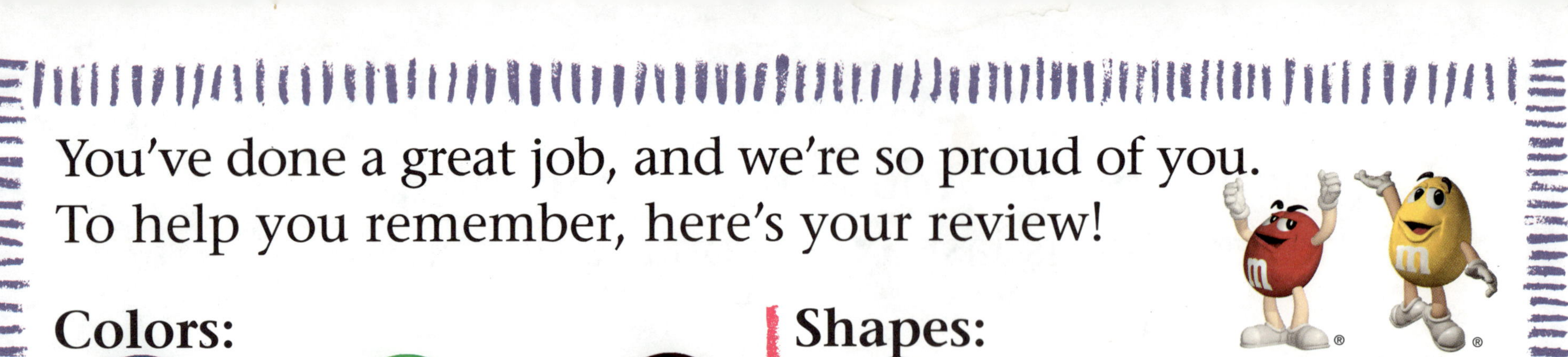

You've done a great job, and we're so proud of you.
To help you remember, here's your review!

Colors:

blue | green | brown

orange | yellow | red

Shapes:

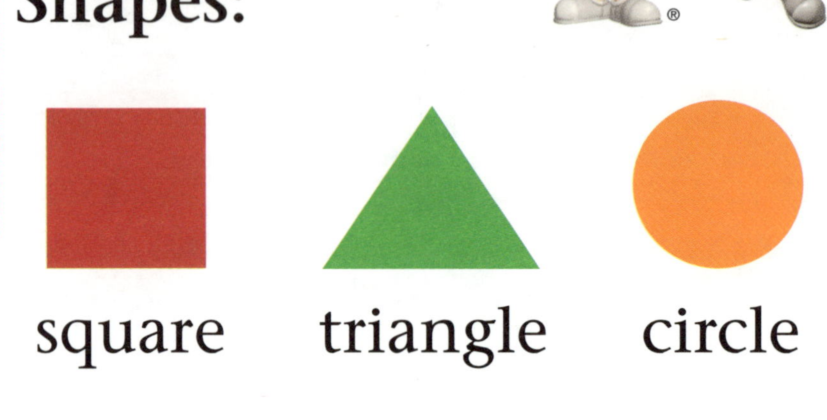

square | triangle | circle

Numbers:

1 One	2 Two	3 Three
4 Four	5 Five	6 Six
7 Seven	8 Eight	9 Nine
10 Ten	11 Eleven	12 Twelve

The Sets of 12:

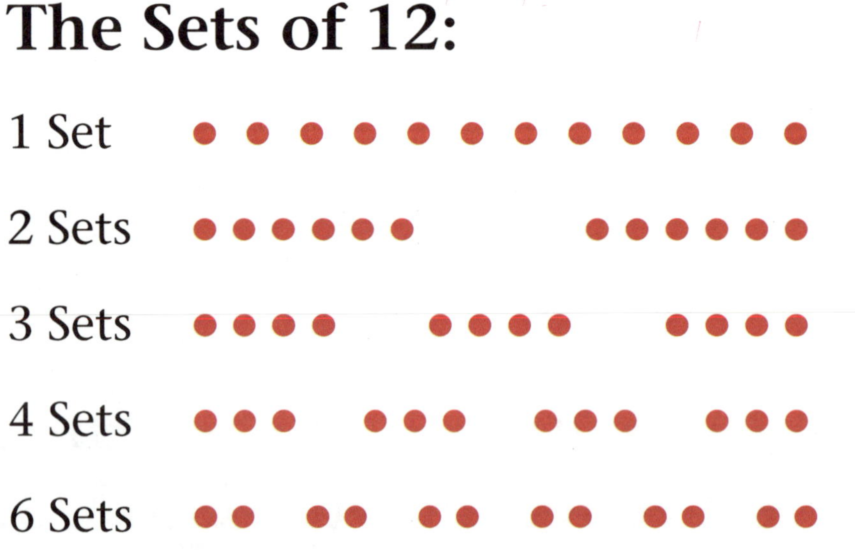